SURVIVAL STORIES

SURVIVING THE RAINFOREST

By Vicki C. Hayes

Kaleidoscope
Minneapolis, MN

BIGFOOT BOOKS

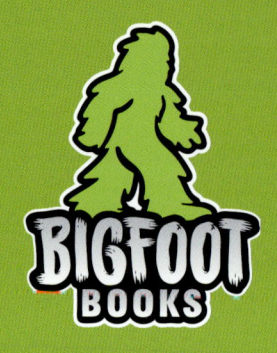

The Quest for Discovery Never Ends

....

This edition is co-published by agreement between
Kaleidoscope and World Book, Inc.

Kaleidoscope Publishing, Inc.
6012 Blue Circle Drive
Minnetonka, MN 55343 U.S.A.

World Book, Inc.
180 North LaSalle St., Suite 900
Chicago IL 60601 U.S.A.

All rights reserved. No part of this book may be reproduced in any form without written permission from the publishers.

Kaleidoscope ISBNs
978-1-64519-207-7 (library bound)
978-1-64519-275-6 (ebook)

World Book ISBN
978-0-7166-4175-9 (library bound)

Library of Congress Control Number
2020936077

Text copyright © 2021 by Kaleidoscope Publishing, Inc. All-Star Sports, Bigfoot Books, and associated logos are trademarks and/or registered trademarks of Kaleidoscope Publishing, Inc.

Developed and produced by Focus Strategic Communications Inc.

Printed in the United States of America.

FIND ME IF YOU CAN!

Bigfoot lurks within one of the images in this book. It's up to you to find him!

TABLE OF CONTENTS

Survival Story 1: Trapped in a Jungle Pool 4

Survival Story 2: Wandering in the Rainforest 10

Survival Story 3: Swept Away in the Jungle 16

Survival Story 4: Two Miracle Survivals 22

Beyond the Book 28
Research Ninja 29
Further Resources 30
Glossary 31
Index 32
Photo Credits 32
About the Author 32

Survival Story 1
Trapped in a Jungle Pool

In 2017, Paul Nicholls was exploring the Khun Si Falls on the island of Koh Samui, Thailand. He was all alone in the rainforest when the wild dogs came. There were 15 of them. They were between him and his rented motorbike.

He wondered where he could go to get away. He looked at the rainforest. It was infested with snakes, spiders, and mosquitoes. He looked at the waterfall. It plunged down almost 20 feet (6 m). He remembered being told that a month earlier a tourist had died after falling from this waterfall.

Then, a dog attacked. It chomped on his toe. Blood spurted out. Paul decided to jump over the waterfall. He landed in a pool. His ribs hurt, his leg felt broken, and his kneecap was smashed. Blood swirled into the water.

FUN FACT
The island of Koh Sumui has many waterfalls. The Na Muang 1 and 2 are the main ones. Khun Si Falls is harder to get to.

Koh Samui is the third largest island in Thailand. It is covered with rainforests.

Wild dogs live and travel in packs. They can be aggressive toward humans.

Many animals live near waterfalls.

Big leeches and crabs started swimming toward Paul. The blood was attracting them. He swatted at the crabs. They retreated. The leeches were tougher. They came out of the mud and holes in the rocks. He batted them away and screamed for help. No one heard him.

All day, Paul battled the leeches. Then, he worried about nightfall. He wondered how he would see them. He dropped down to another pool of water about 15 feet (4.5 m) lower. The pain was horrible. He could not move out of the water. He could not find safety or food. He became **delirious** and semi-conscious. He had **hallucinations**.

WHAT ARE LEECHES?

A leech is a kind of worm that feeds on blood. Leeches use their suckers to attach themselves onto a living being. Then, they bite and suck out the being's blood.

A rescue team made up of military and police searched for Paul.

Meanwhile, a villager found Paul's motorbike. The police were notified. A search began. Rescuers hacked through the rainforest undergrowth. When they reached Paul, he burst into tears. It had been three days. He had been in the water with no food, a broken leg, and a missing kneecap. He had a tropical infection from his untreated wounds. He would not have survived much longer.

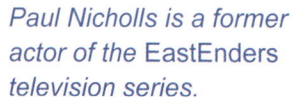

Paul Nicholls is a former actor of the EastEnders *television series.*

RAINFORESTS OF THE WORLD

Arctic Ocean
GREENLAND
UK
EUROPE
NORTH AMERICA
Atlantic Ocean
AFRICA
ASIA
Pacific Ocean
Indian Ocean
AUSTRALIA
SOUTH AMERICA
NEW ZEALAND

= Rainforests

WHAT ARE RAINFORESTS?

Rainforests are found all over the world. They are dense jungles with tall trees and lots of rain. The largest is the Amazon rainforest in South America.

Survival Story 2
Wandering in the Rainforest

On May 8, 2019, Amanda Eller went for a run at the Maui Makawao Forest Reserve in Hawaii. Now, she was lost. She was trying to get back to her car. It had her cell phone and water bottle inside. She had been

The Makawao Forest Reserve is on the island of Maui in Hawaii.

walking in the Hawaiian jungle for hours. But she had gone in the wrong direction.

Soon, Amanda became hungry and thirsty. She ate some wild berries and guava. She drank clear river water. Days passed. She was getting too thin. She ate moths that landed on her body.

Guava are very nutritious. They contain four times more vitamin C than oranges.

Wild boars make their dens out of sticks and branches.

FUN FACT

Wild boars can be very dangerous to humans. They are very powerful. They also have sharp tusks and claws.

The days were hot, but the nights were cold. Amanda covered herself with ferns and leaves to stay warm. Sometimes, she slept in the mud. One night, she slept in a wild boar's den. She fought her way through the dense forest. There were steep **ravines**, sharp lava rocks, and giant ferns.

Rainforests have very rugged terrain.

One day, Amanda lost her shoes in a flash flood. Her feet and ankles became covered with infections. Then, she slipped off a 20-foot (6-m) cliff. She broke her leg. She tore her knee. Now, she had to crawl. She became severely sunburned.

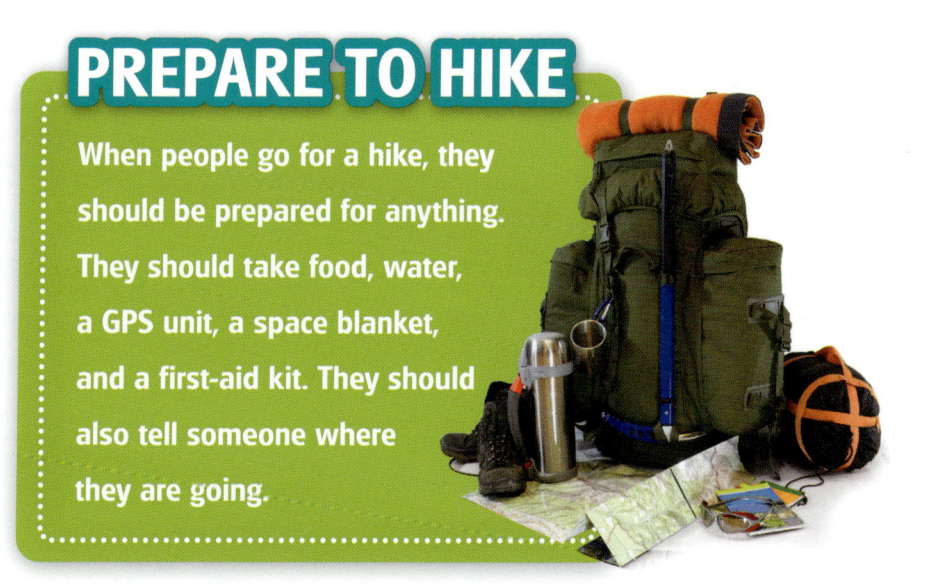

PREPARE TO HIKE

When people go for a hike, they should be prepared for anything. They should take food, water, a GPS unit, a space blanket, and a first-aid kit. They should also tell someone where they are going.

Waterfalls are pretty, but they can be dangerous if climbed.

Amanda wondered if rescuers were looking for her. They were. Thousands of volunteers were scouring the jungle. They **rappelled** into ravines. They searched caves. They dove into pools. They even killed wild boars and checked their stomachs for human remains. But they were looking too close to Amanda's car.

Then on day 17, a helicopter crew decided to fly farther out. They found Amanda in a steep canyon with waterfalls on both sides. They managed to lift her out. She had walked and crawled more than 30 miles (50 km), but now she was saved.

FUN FACT

Rescue helicopters can hover, fly up and down, and fit into tight spaces. This makes it possible to rescue people in hard conditions.

Amanda smiles with her rescuers.

Survival Story 3
Swept Away in the Jungle

A week before, Yossi Ghinsberg had been so excited about **trekking** through the Amazon rainforest in Bolivia. It was 1981. Yossi and his friend were on a river raft. But they hit dangerous rapids and lost control. Yossi was swept over a waterfall and into a deep canyon. He struggled back upriver looking for his friend but found no one.

An aerial view of a rainforest makes the trees look like they go on forever.

Black capped squirrel monkeys are found in Bolivia, Brazil, and Peru.

Days passed. The noise from the jungle creatures was deafening. Yossi was hungry. He looked for berries and wild fruits. He grabbed eggs from nests. Once, he waited for a monkey to fall so he could eat it.

Fire ant bites burn.

FUN FACT
The biggest threats in a rainforest are not the large animals. The biggest threats are the millions of insects.

The rainforest was full of scorpions, spiders, and poisonous snakes. One night, Yossi was attacked by termites that almost ate him alive. Two nights later, he was covered with stinging fire ants. Blood-sucking worms burrowed into his skin. He dug 14 out of his forehead.

Yossi met a wild boar with razor-sharp claws. Once, he woke up with a jaguar sniffing him. He grabbed a can of mosquito repellent and a lighter from his backpack. He made a flame-thrower to scare it away.

Jaguars live in rainforests. There are about 15,000 jaguars left in the wild.

AMAZON RAINFOREST
ANIMALS

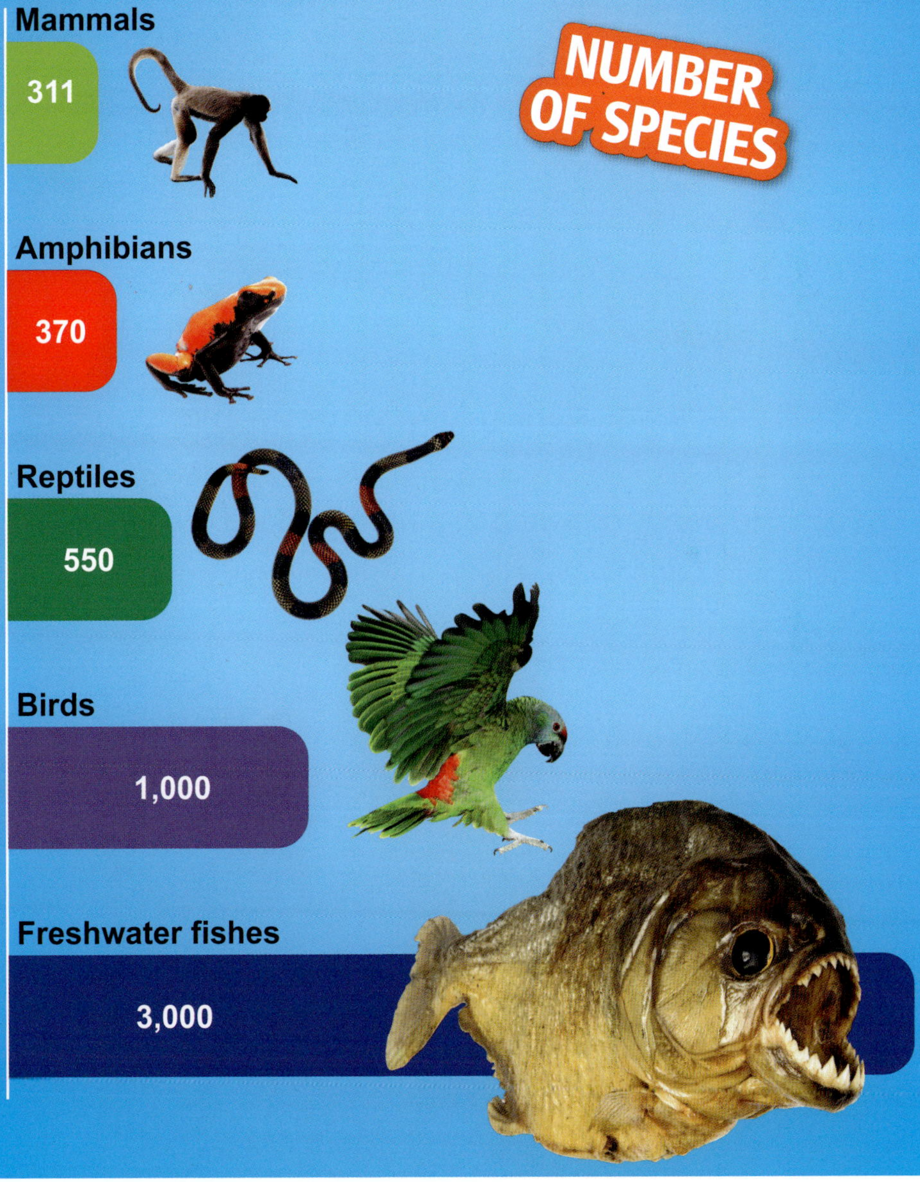

NUMBER OF SPECIES

Mammals
311

Amphibians
370

Reptiles
550

Birds
1,000

Freshwater fishes
3,000

Yossi's friends searched for him by boat.

Yossi's body was starting to waste away. The skin came off his feet. He was walking on chunks of flesh. He nearly drowned in a horrible flood. He sank into bogs twice. His skin hung on his bones. He had hallucinations.

Yossi Ghinsberg is an author and motivational speaker. He shares his Amazon experience with others.

Then, after 20 days in the jungle, Yossi heard an engine. He went to the river and saw a boat. His friends had been looking for him. He was too exhausted to move or yell. But his friend turned and saw him. Yossi was rescued.

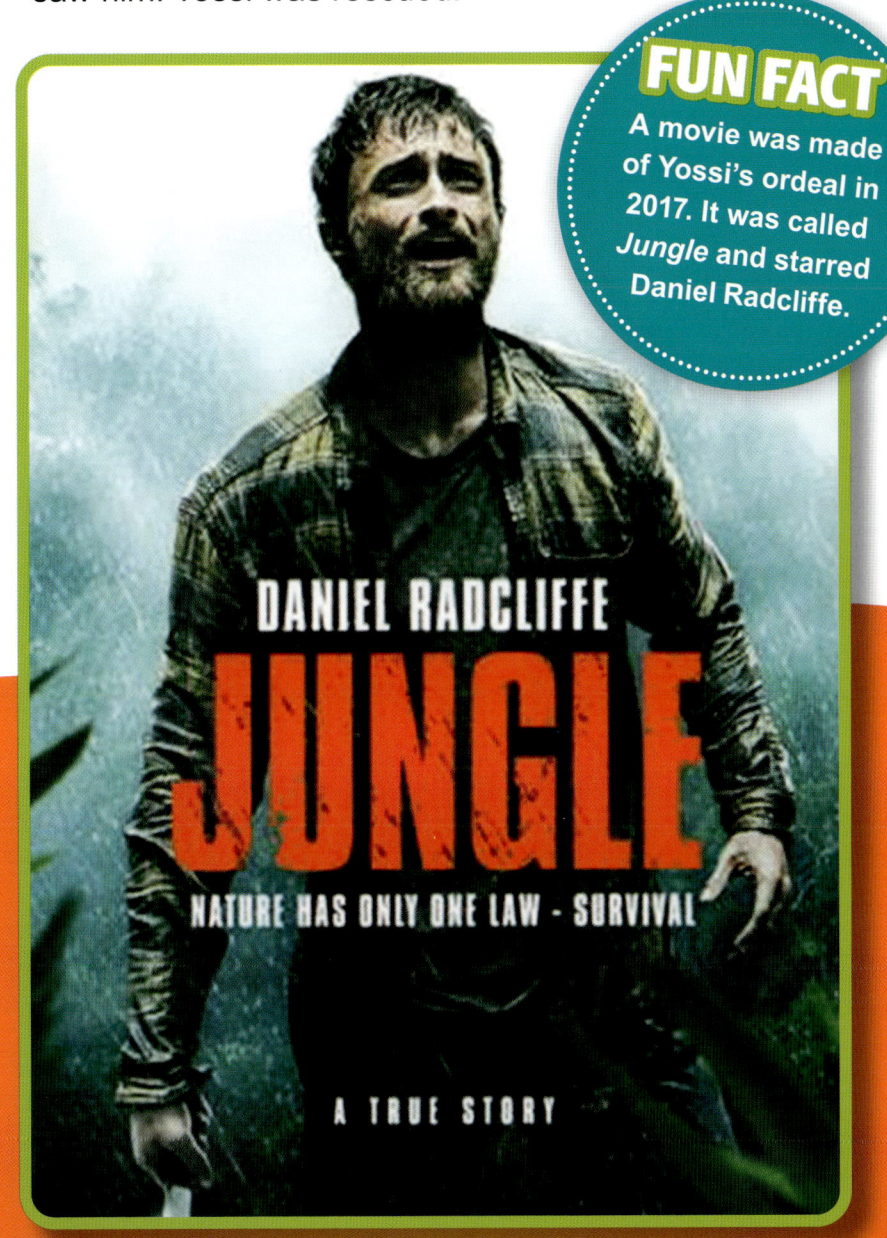

FUN FACT
A movie was made of Yossi's ordeal in 2017. It was called *Jungle* and starred Daniel Radcliffe.

Survival Story 4
Two Miracle Survivals

It was December 24, 1971. Luggage fell out of the overhead lockers. Food and drinks flew around the cabin. Seventeen-year-old Juliane Koepcke grabbed her mother's hand. The heavy **turbulence** was awful. The plane had been hit by lightning. It went into a nosedive. Everything went pitch black. The roar of the engines blocked out the sounds of people crying and screaming. Then, it was silent.

Juliane opened her eyes. She was in freefall outside the plane. She was still strapped to her seat at 10,000 feet (3050 m) in the air. The tree canopy below was approaching. She lost consciousness.

Planes do not crash often, but lightning can sometimes bring them down.

Water in the rainforest can be full of creepy creatures.

24

When Juliane awoke, she was on the ground. She had survived a plane crash with only a broken collarbone and a few gashes. She looked for her mother, but there were no other survivors.

Juliane followed a tiny jungle stream. When it got bigger, she walked in the water. She tried to avoid stingrays, **piranhas**, and alligators. She heard a vulture. She knew it was eating bodies from the crash.

FUN FACT
Vultures eat fresh or rotten meat. They help the environment because they clean up corpses that could spread disease.

Incredible true-life drama of Juliane Koepcke... the story of a 17 year old girl who survived a 10,000 foot plunge from an exploding jet and an 11 day terrifying ordeal in the Peruvian jungle!

"Miracles Still Happen"

A BRUT PRODUCTION • Starring SUSAN PENHALIGON • GRAZIELLA GALVANI • PAUL MULLER
Story and Screenplay by GIUSEPPE SCOTESE • Produced by NINKI MASLANSKY • Directed by GIUSEPPE SCOTESE
Music by MARCELLO GIOMBINI • Color by TECHNICOLOR • From BRUT Film Distributors, Inc.

Juliane walked in the water for days. The second-degree sunburn on her back started bleeding. Her wounds became infested with **maggots**. She was starving and weak. She started hallucinating.

Then, Juliane saw a small boat. She sucked gasoline out of the fuel tank and put it on her arm. The maggots burrowed deeper. The pain was terrible. Then, the maggots suffocated. She pulled out 30. She fell asleep exhausted. The next morning, a local fisherman found her. After 10 days, she was saved.

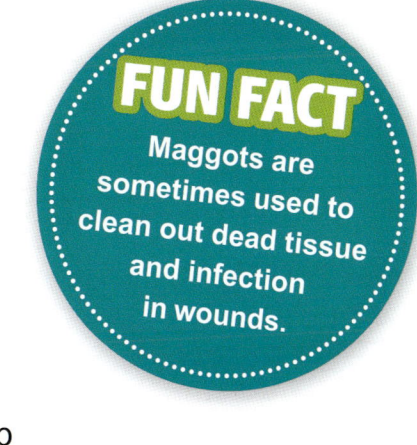

FUN FACT
Maggots are sometimes used to clean out dead tissue and infection in wounds.

Juliane poses with her autobiography.

BEYOND
THE BOOK

After reading the book, it's time to think about what you learned. Try the following exercises to jumpstart your ideas.

THINK

DIFFERENT SOURCES. Think about what types of sources you could find about rainforests. What could you find in an encyclopedia? What could you learn in other books? How could each of the sources be useful in its own way?

CREATE

PRIMARY SOURCES. A primary source is an original document, photograph, or interview. Make a list of primary sources you might be able to find about rainforests. What new information might you learn from these sources?

SHARE

WHAT'S YOUR OPINION? Rainforests are essential to life on Earth. They provide air, water, medicine, and food. They also absorb greenhouse gases. Some people think rainforests should be left alone and protected. Others feel they can be carefully explored. What do you think? Use evidence from the text to support your answer. Share your position and evidence with a friend. Does your friend agree with you?

GROW

REAL-LIFE RESEARCH. Think about what kinds of places you could visit to learn more about rainforests. What other topics could you explore there?

RESEARCH NINJA

Visit **www.ninjaresearcher.com/2077** to learn how to take your research skills and book report writing to the next level!

RESEARCH

DIGITAL LITERACY TOOLS

SEARCH LIKE A PRO
Learn how to use search engines to find useful websites.

FACT OR FAKE?
Discover how you can tell a trusted website from an untrustworthy resource.

TEXT DETECTIVE
Explore how to zero in on the information you need most.

SHOW YOUR WORK
Research responsibly—learn how to cite sources.

WRITE

GET TO THE POINT
Learn how to express your main ideas.

PLAN OF ATTACK
Learn prewriting exercises and create an outline.

DOWNLOADABLE REPORT FORMS

Further Resources

BOOKS

Fabiny, Sarah. *Where is the Amazon?* New York, NY: Who HQ, 2016.

Franchino, Vicky. *Amazon Rain Forest*. North Mankato, MN: Cherry Lake Publishing, 2016.

Pearl, Melissa Sherman. *Kids Saving the Rainforest*. North Mankato, MN: Cherry Lake Publishing, 2018.

Vonder Brink, Tracy. *Protecting the Amazon Rainforest*. Mendota Heights, MN: Focus Readers, 2020.

WEBSITES

FACTSURFER

Factsurfer.com gives you a safe, fun way to find more information.

1. Go to www.factsurfer.com.
2. Enter "Surviving the Rainforest" into the search box and click 🔍
3. Select your book cover to see a list of related websites.

Glossary

delirious: Mentally confused due to severe conditions. Prolonged stress, extreme fatigue, excessive heat or cold, or starvation can all make a person delirious.

hallucinations: Something a person sees, but it is not actually there. Severe illnesses or other very stressful situations can cause hallucinations.

maggots: Fly larvae. When flies lay their eggs in damaged human tissue, larvae develop and feed off the decaying tissue.

piranhas: Fish with razor-sharp teeth. Piranhas have lived in the Amazon River for millions of years.

rappelled: Moved down a vertical rock using ropes. Climbers first attach their ropes to a rock, and then they rappell down the steep cliff.

ravines: Narrow canyons with steep sides. Ravines form when a stream erodes the land.

trekking: Going on a long, usually difficult journey. Trekking through the jungle can be dangerous.

turbulence: The violent movement of water or air. Air turbulence is usually caused by hot air rising quickly, and it can make riding in a plane very uncomfortable.

Index

Amazon Rainforest, 9, 16, 19, 20
Bolivia, 16, 17
EastEnders, 8
Eller, Amanda, 10, 11, 12, 13, 14, 15
fire ants, 18
Ghinsberg, Yossi, 16, 17, 18, 20, 21
hallucination, 7, 20, 26
Hawaii, 10, 11
helicopter, 15
Jungle, 21
Koepcke, Juliane, 22, 25, 26, 27
leeches, 6, 7
Miracles Still Happen, 26
Nicholls, Paul, 4, 6, 7, 8
piranhas, 25
Radcliffe, Daniel, 21
Thailand, 4, 5
wild boars, 12, 14, 18
wild dogs, 4, 5

PHOTO CREDITS

The images in this book are reproduced through the courtesy of: Quick Shot/Shutterstock Images, front cover (top), p. 1 (top); Ivana Casanova/Shutterstock Images, front cover (skulls); AustralianCamera/Shutterstock Images, front cover (bottom), p. 1 (bottom); Dzmitrock/Shutterstock Images, front cover (hiker); Ilona Marienkova/Shutterstock Images, p. 5 (top); paula french/Shutterstock Images, p. 5 (bottom); JuliiaKosh/Shutterstock Images, p. 6 (top); Pixfiction/Shutterstock Images, p. 6 (bottom); frank60/Shutterstock Images, p. 7; umarazak/Shutterstock Images, p. 8 (top); Tim Whitby/PA Wire URN:32111391/AP Images, p. 8 (bottom); lukmanhakim/Shutterstock Images, p. 9 (compass icon); George Burba/Shutterstock Images, pp. 10-11 (bottom); Writefully Said/Shutterstock Images, p. 11; Kirsanov Valeriy Vladimirovich/Shutterstock Images, p. 12 (top); Eric Isselee/Shutterstock Images, p. 12 (wild boar); Uwe Bergwitz/Shutterstock Images, p. 13 (top); IgorXIII/Shutterstock Images, p. 13 (bottom); Tatiana Nurieva/Shutterstock Images, p. 14; JAVIER CANTELLOPS/AFP/Getty Images, p. 15; Gustavo Frazao/Shutterstock Images, pp. 16-17; Jolanda Aalbers/Shutterstock Images, p. 17; PongMoji/Shutterstock Images, p. 18 (top); Adalbert Dragon/Shutterstock Images, p. 18 (bottom); Iakov Filimonov/Shutterstock Images, p. 19 (monkey); Dirk Ercken/Shutterstock Images, p. 19 (frog); Rosa Jay/Shutterstock Images, p. 19 (parrot); Dr. Morley Read/Shutterstock Images, p. 19 (snake); Andrew Burgess/Shutterstock Images, p. 19 (piranha); RPBaiao/Shutterstock Images, p. 20 (top); Narwhalco/CC BY-SA 4.0, p. 20 (bottom); Ysanne Slide/Moviestore Collection Ltd/Alamy Stock Photo, p. 21; Denis Belitsky/Shutterstock Images, pp. 22-23; simongee/Shutterstock Images, p. 24 (top); Kent Ellington/Shutterstock Images, p. 24 (bottom); Dennis Jacobsen/Shutterstock Images, p. 25; LMPC/Getty Images, p. 26 (top); StockMediaSeller/Shutterstock Images, p. 26 (bottom); Martin Acevedo/GDA/AP Images, p. 27.

About the Author

Vicki C. Hayes lives and works Seattle with her husband, her dog, and four nearby grandchildren. She loves hiking on forest trails but is careful not to get lost.